THE ANNALS OF
THE KING'S ROYAL RIFLE CORPS

APPENDIX

DEALING WITH

UNIFORM, ARMAMENT AND EQUIPMENT

LIEUT-GENERAL SIR FREDERICK HALDIMAND, K.B.,
GOVERNOR GENERAL OF CANADA.
Lieut.-Col. 60th Regt., 1756—72.
Col. Commdt. ,, 1772—91.

THE ANNALS OF THE KING'S ROYAL RIFLE CORPS

APPENDIX

DEALING WITH

UNIFORM, ARMAMENT AND EQUIPMENT

BY

S. M. MILNE

AND

MAJOR-GENERAL ASTLEY TERRY

LIST OF PLATES

[*Coloured Plates are marked* †]

Plate A.

Officer,
Grenadier Companies,
1755.

Private,
Battalion Companies,
1758.

Plate C.

Private.

Grenadier Companies.

1768.

Officer,
Battalion Companies.
1770.

Plate E.

Officer.
Battalion Companies.
1792.

Officer & Private of a Light Company.
1801.

Serjeant, Grenadier Company, 2nd Battalion,
1801.
[From a German print.]

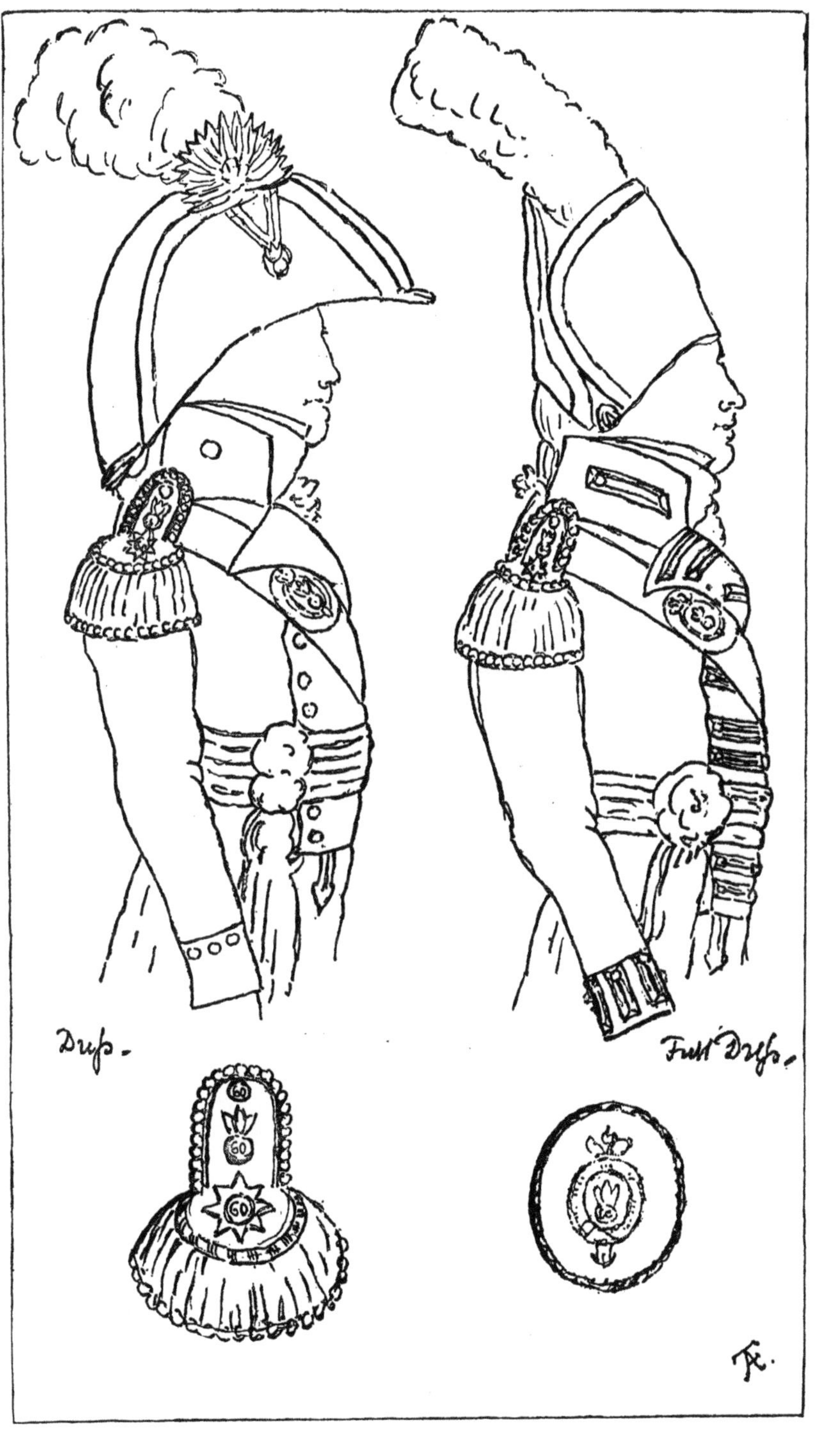

Officer, Grenadier Company, 4th Battalion,
1801.
[From a miniature of Lieut. George Bent.]

Plate H*

Officer, 5th Rifle Battalion, 1797

From a sketch of Lieut. John Anthony Wolff.

(Lieut. 30th Dec. 1797, Capt 5th June 1806, wounded at Talavera, Half Pay 1815.)

5th Battalion. 1797.　　6th Battalion. 1799.

Riflemen.

From drawings by Col. C. Hamilton Smith, in the S.K. Museum.

Plate K.

Rifleman, 5th Battalion.

1812.

Taken from a print after Col. Hamilton Smith, D.A.Q.M.G.

Officer, 5th Battalion.
1812.
From an old drawing.

Officer, 1st (Light Infantry) Battalion.
Full Dress,
1815.
From contemporary records.

Officer, 7th (Light Infantry) Battalion

Dress.

1817.

From an old drawing.

Plate O

Field Officer, 1st Battalion.
1824-27.

From an oil painting by Heath, in the possession of Maj. Gen. Astley Terry.

Plate Q.

Serjeant

1824–30.

From a print after E. Hull.

Officer, 1st Battalion,

1830–34.

From a print after Mansion and St Eschauzier.

Serjeant and Private Rifleman,

1842.

From an oil painting by Collins, in the possession of Maj. Gen. Astley Terry.

Officer, Canadian Dress.

1846.

From a print after H. Martens.

Plate V.

Corporal and Private Rifleman.
1852.

Bugler and Bandsman.

1854.

Plate B.B

Rifleman.
1858.

Officer and Corporal.
Marching Order.
1873–78.

Plate D.D.

Officers, Mess Dress
till 1902.

Plate E.E.

Serjeant-Major and Bugler.
1880.

Field Officer, 1st Battalion.

1885.

Officer *Officer and Private Rifleman.*

Undress. *Review Order.*

1904.

King's Colour.

Colours of the First Battalion.
1802–1816.

PLATE J.

60th Rgt   95th Rgt

BRITISH RIFLEMEN.

1812.

[From a print after Col. Chas. Hamilton Smith, D.A.Q.M.G.]

Plate M.*

Private,
1st (Light Infantry) Battalion,
1816.
[From a drawing by Collins, in possession of Major-Genl Astley Terry.]

Plate N*

H.R.H. Frederick, Duke of York, K.G.,
Colonel-in-Chief, 1797—1827.

PLATE P.

Group of Officers
(with target practice),
1828.
[From a print after W. Heath.]

Field Officer,
1830-34.

Officer,
1834-5.
[From a print after Mansion and St. Eschauzier.]

Non-Commissioned Officers and Privates,
1840.
[From a dated painting by M. A. Hayes, in the possession of Major-Genl. Astley Terry.]

2nd Battalion, 1850.
[From a painting by M. A. Hayes, in the possession of Major-Genl. Astley Terry.]

PLATE AA.

Bandsman, 1856.

PLATE a.

The Colours, 1755.

Silver Gorget.
Worn by Lieutenant John Johnes,
Who was appointed to the 4th Battn. 25th September,
1787.

Fig. 1.
Soldier's Pewter Button,
(Slightly domed)
1767—1782.

Fig 2
Officer's Silver Button.
Soldier's Pewter
(Flat)
1782—1799.

Fig. 3
Officer's silver button,
(domed)
1799 – 1815.

Fig. 4.
Officer's
Engraved Silver Breast-plate,
1800 (or earlier)—1812.

Fig. 1. Soldier's
Brass—Embossed,
1800.

Fig. 2. Officer's
Silver—Engraved,
1812—15.

Breast-plates.

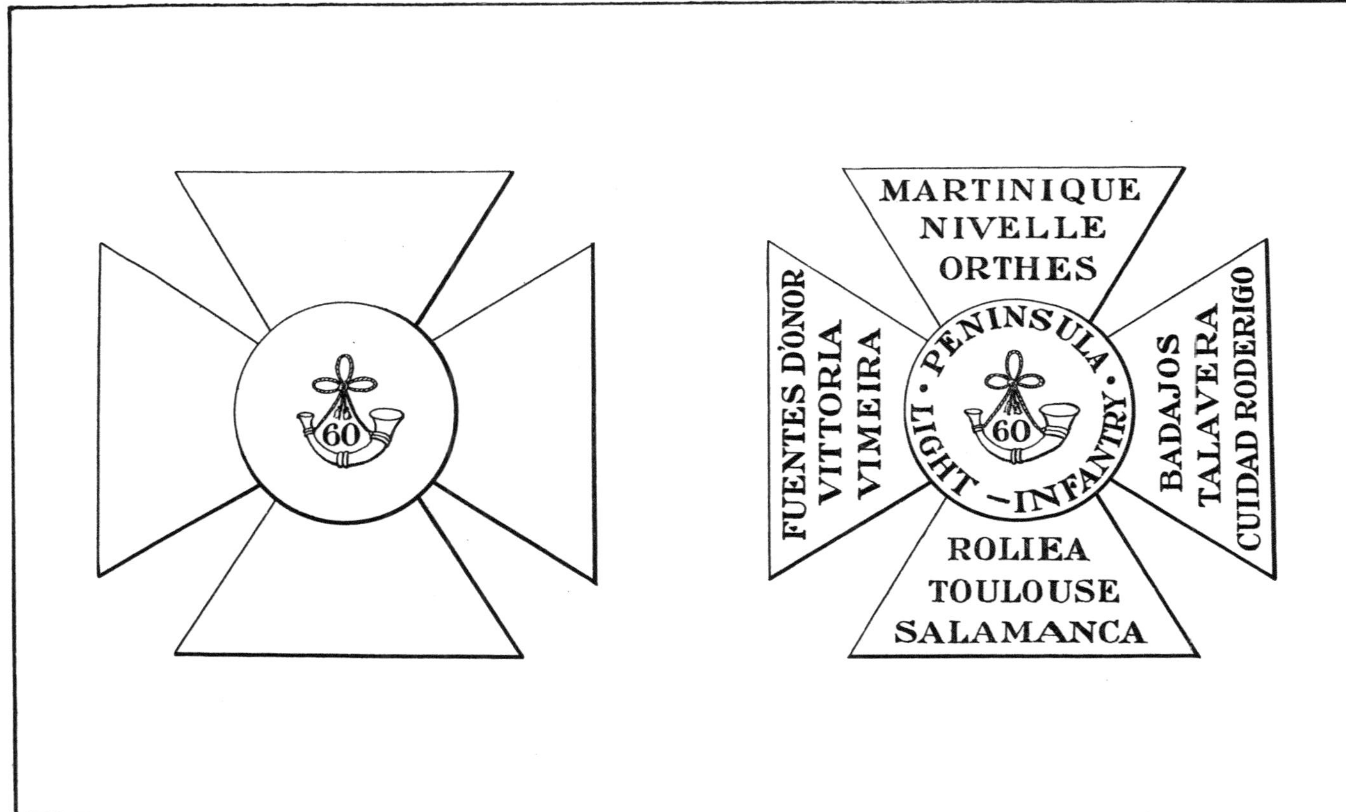

Fig. 1. 1797—1818 (5th Battn.)
1815—1821 (other Battns.)

Fig. 2. 1121—22 (2nd Battn.)

Silver Cross on Pouch-belt.

Silver Ornaments on Pouch-belt,
1822–27.
[Scroll worn by 1st Batt. from 1824.]

Fig. 1.
1827–83.
(Title changed in 1830,
and Honours added as authorized.)

Fig. 2.
From 1827.

Fig 3.
1883–1906.
Also in bronze with perforated centre, on helmet,
1883–90.

Silver Ornaments on Pouch-belt.

'The Peninsula Cross'
from 1906.
The present badge of the Regiment.

Bronze Chaco Plate, 1830—34.
[Title changed December, 1830.]
(Without crown and star 1834—5, but with small crown on cockade.)

Fig. 1.
On Chaco, 1835—61.
(Smaller on chaco and busby, 1861—78.)

Fig. 2.
On Helmet, 1878—83.
(Scarlet cloth under perforated centre.)

Bronze Cap Plates.

PLATE k.

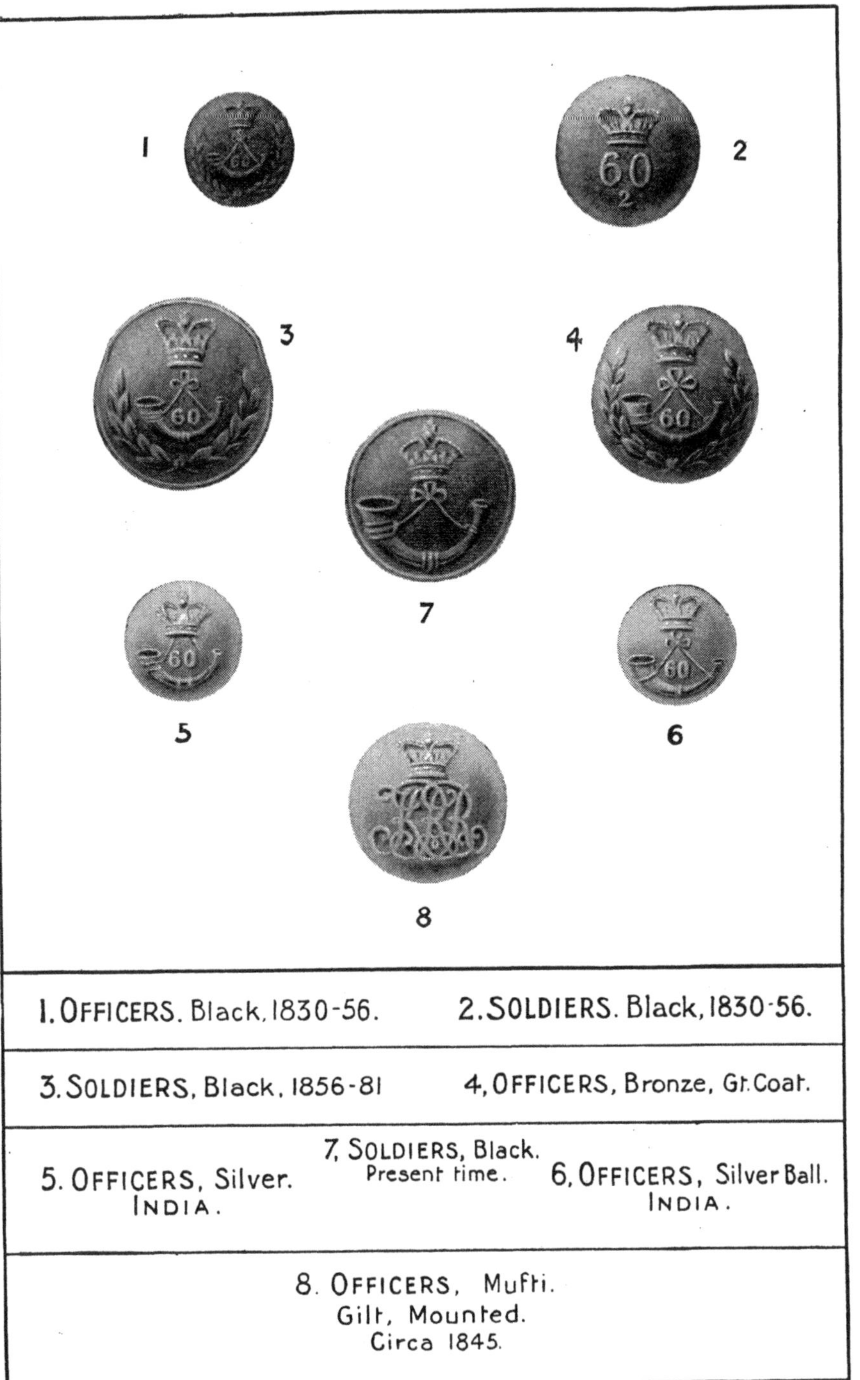

BUTTONS.

[The "60" was discontinued in 1881.]

Regimental Medals.

1. Large Silver Medal.

2. Small Gold Medal.

3. Bronze Cross.

THE DRESS OF THE REGIMENT

PART I

By S. M. MILNE

'THE RED COAT'

The dress and equipment of the 60th Regiment raised in 1755 were in exact conformity with the clothing regulations of the British Army, save and except that the private soldiers wore no regimental lace.[1] Generally for the rest of the infantry white worsted braid with coloured stripes or markings was worn upon the coat, and this lace, together with the different coloured facings, served, in the absence of numbers upon the buttons, to distinguish one regiment from another.

The following letter bears upon the subject of lace:

'London, November 6, 1758.

'Application having been made for permission to make the clothing of the regiments serving in America without lace, His Lordship does not think himself at liberty to dispense with His Majesty's orders, which direct the soldiers' coats to be laced, excepting the Regiment of Royal Americans, which from its first raising

[1] This wise provision was made with regard to the rough and wooded country in which the regiment was destined to operate.

was permitted to be without lace, also Gage's Regiment of Light Infantry.

'ROBERT NAPIER,
'Adj.-Genl.'

The coloured plate 'A' represents an officer of a grenadier company, just after the regiment was raised. His grenadier cap, probably of blue velvet, was richly embroidered with silver and gold thread, and bore in front the device of the White Horse, together with the words 'Nec aspera terrent,' being the badge and motto of the reigning House of Hanover, common, however, to all grenadier officers and privates throughout the service. The silver aiguillette upon the right shoulder, together with the crimson silk sash, and the silver gorget, denote the officer. He carries a fusee or fusil, a light rifled musket.

The grenadier company had two pipers, besides the drummer, the only company which had these piercing instruments.

The next illustration, Plate 'B,' is that of a private soldier, without any lace upon the blue facings of his coat, and, as a consequence, looking quite plainly dressed compared with the soldiers of other regiments.

A royal warrant was issued September 21, 1767, ordering the regimental number to appear upon the coat buttons of both officers and men; hitherto these had been without device, and quite plain.

The illustration, Plate 'c,' fig. 1, represents a private soldier's button found not very long ago at Fort George, Lake Champlain, North America, and no doubt lost there at some period before 1772—and probably some years earlier. The button is made of pewter, and has the number '60' deeply engraved in the centre, the edge ornamented by a band of rope pattern, also engraved.

At length, after a long period of irregularities, the government issued a royal warrant, dated December 19, 1768, laying down with much distinctness the details of dress of both officers and men of the infantry, which applied to all regiments, the 60th included.

Officers' coats, lapelled to the waist with blue cloth—these might be without embroidery or lace if required—to have cross pockets, and sleeves with round cuffs, and no slits. The lapels and cuffs to be the same breadth as those for the men.

Officers of the 1st or grenadier company to wear an epaulette on each shoulder. Those of the battalion to wear one upon the right shoulder. They were to be either of lace or embroidery, those of the 60th with silver fringe. Waistcoats to be plain white cloth, without lace.

Officers' swords to be uniform, and sword-knots to be of crimson and gold in stripes. The hilt of the swords of the 60th to be of silver, according to the colour of the buttons of the uniform.

Hats to be laced with silver, and to be uniformly cocked. Sashes to be of crimson silk, and worn round the waist.

The King's arms to be engraved on the gorgets; also the number of the regiment. The gorgets to be of silver.

Officers of the grenadier company to wear black bearskin caps, and to have fusils, white shoulder-belts, and pouches. Battalion officers to have espontoons. The whole to have black linen gaiters with black buttons and small stiff tops, black garters, and uniform buckles.

Serjeants' coats to be lapelled to the waist with blue, the buttonholes to be of white braid. Serjeants of grenadier companies to have swords, fusils, pouches, and caps; those of the battalion to have swords and halberts only.

Sashes to be of crimson worsted, with a stripe of blue, and worn round the waist.

Corporals' coats to have a silk epaulette on the right shoulder. Grenadiers' coats to have the usual round wings of red cloth on the point of the shoulder, with six loops of the regimental lace (white worsted with a blue stripe) and a border of the same round the bottom.

Private men's coats to be looped with worsted lace, but no border, the ground of the lace to be white with one blue stripe; to have white buttons; four loops to be on the sleeves, and four on the pockets, with two on the slit behind. The breadth of all the lapels to be three inches, to reach down to the waist, and not to be wider at the top than at the bottom. The sleeves of the coats to have a small round cuff without any slit, and to be made so that they may be unbuttoned and let down. The whole to have cross pockets, but no flaps to those of the waistcoat. The cuff of the sleeve, which turns up, to be three inches and a half deep. The flap on the pockets of the coat to be sewed down, and the pocket to be cut in the lining of the coat.

Drummers' and fifers' coats to be red, faced and lapelled with blue; waistcoats, breeches and linings to be white; to be laced as the colonel thinks fit, the lace being of the regimental pattern. On the front of the drummers' and fifers' bearskin caps the King's crest in silver-plated metal on a black ground, with trophies of colours and drums; the number of the regiment on the back part.

The grenadiers also wore the King's crest on their bearskin caps, but with the motto 'Nec aspera terrent' and a grenade on the back part with the number of the regiment on it.

Hats of the serjeants to be laced with silver, those of

the corporals and private men with white tape; all hats to have black cockades.

Each pioneer to have an axe, a saw, and an apron; a cap with a leather crown and black bearskin front, on which is to be the King's crest in white on a red ground, also an axe and a saw; the number of the regiment to be on the back part of the cap. Hat lace for the officers, silver; waistcoat, breeches, and lining of coats, white.

The coloured illustration, Plate 'C,' represents a grenadier of the regiment, drawn in accordance with the new regulations. The men of the battalion companies wore the same coat without the shoulder wings, and a three-cornered cocked hat, ornamented with white tape lace. The brass match-case will be observed fastened upon the shoulder-belt; formerly this case contained the lighted slow-match for the purpose of igniting the hand-grenades which were thrown by the men. It remained simply a mark of distinction for the men of the grenadier company, no grenades having been used for a considerable period. This plate has been copied from a drawing in an old book of uniforms in the library at Aldershot, dated 1769.

The lace upon the coats had now become an object of importance in the army and a matter of strict regulation, as the following letter from headquarters, copied from the General Officers' Letter Books, Record Office, Chancery Lane, will testify:

'Cleveland Court, September 26, 1768.

'I am desired to acquaint the several clothiers that as soon as each Colonel of infantry has fixed upon the patterns of woosted [*sic*] lace which he proposes to be for the looping of the next clothing, a piece of it of the length of half a yard is to be left at the Comptroller's office in order that it may be shewn to His Majesty. If His Majesty approves of it, orders will be given that no alteration be made for the future either in the colour of the ground, stripes, or breadth. A piece of the same cloth of which the lappett for ye pattern coat of each regiment is made, is also to be left at the same time.

'E. HARVEY,
'Adjt.-Genl.

'T. Fanquier Esquire.'

It may be added that very many different patterns of lace were in use which, with the numerous hues and shades of the facing cloth, served to distinguish one regiment from another. The coloured illustration, Plate 'D,' represents an officer of the battalion companies in full dress, also in accordance with the new regulations. The quantity of silver lace upon the coat had been considerably reduced, and the sword-belt worn under the coat; crimson sash, now tied round the waist, a silver buckle and tip just showing, the forerunner of the silver breastplate soon to be introduced; a small silver epaulette has taken the place of the silver aiguillette, so long the distinguishing badge of the officer; he carries also the espontoon, a weapon used with graceful effect in the salute. The gorget is silver, suspended by blue silk ribbons and rosettes.

Light Companies, September 3, 1771.—Orders were issued that a newly-raised light company should be attached to every battalion of infantry, consisting of two serjeants, three corporals, one drummer, and thirty-eight private men. At first, and for many years afterwards, these new companies were clothed in red jackets with very short tails.

About 1782 a new pattern button was introduced. It is believed that, so far, only one specimen of an officer's button has been found, viz. in the United States (see illustration, Plate 'c,' fig. 2). It is quite flat, and made of silver, having '60' stamped in relief in the centre surrounded by a neat border of laurel leaves, springing from a curved stalk also in high relief; the men had a similar button in pewter.

Gorgets.—It is probable that the elaborately ornamented silver gorget (*vide* Plate 'b') was introduced about 1780 or possibly even earlier; it was suspended from two blue silk ribbons terminating in blue rosettes from the button each side of the officer's neck. Probably first introduced into the English service by our Hanoverian sovereigns, it had been in use for a very long time by the kings and princes of reigning German Houses as a personal ornament. A considerable number of these latter may now be seen in the 'Zeughaus,' the military museum in Berlin. It afterwards became the distinguishing badge of the officer, and marked his rank as such. This particular gorget of the 60th Regiment was worn by Lieutenant Johnes, who, however, only served in the regiment about one year, namely 1788, and then exchanged back into the 64th Foot; consequently the gorget itself, having been little worn and evidently well cared for since, still maintains its almost pristine condition. From the Johnes family papers we know that in the year 1788 it cost the owner £2 10*s.*, also that he paid for a regimental coat with silver buttons, lace, and a silver epaulette the sum of £7 19*s.* at the same time. The coat of arms engraved upon the gorget is that used by George III at that period; the regimental number, in Roman characters, is engraved immediately above the embossed trophies of weapons, arms, drums, &c., on the two arms or sides. As we shall find later on, silver gorgets were entirely abolished in 1796.

About 1776 or 1780 the first silver shoulder-belt plate made its appearance in place of the small silver buckle we have seen worn in 1770. The fact that the belt supporting the sword was now worn over, not under, the coat, may have assisted the change by giving a larger surface for ornamentation and for displaying the regimental number, which was engraved thereon. The precise pattern of the earliest plate is not known, but it was probably of an oval shape, somewhat similar in design to that shown in Plate 'c,' fig. 4, which latter was in use about 1799 or 1800. The men wore a brass ornamental plate in the centre of their cross-belts.

In April 1786 orders were issued that battalion officers should use swords instead of espontoons. Both officers and men in general, when under arms or on duty (the grenadier and light infantry companies, when they wore their *caps*, excepted), were, for the future, to wear their hair 'clubbed,' the non-commissioned officers and men to have a small piece of black polished leather, by way of ornament, upon the club. The whole to wear black leather stocks.

On December 10, 1791, it was ordered that the effective field officers of all regiments should in future be distinguished by wearing an epaulette on each shoulder. The officers of the flank companies, who already wore two epaulettes, were to have the addition of a grenade or bugle horn embroidered on each.

The illustration, Plate 'E,' represents an officer of the battalion companies in 1792. Certain small changes may be noticed as having taken place since the date of the last coloured plate, namely that of 1770: the shoulder-belt, outside the coat, and also ornamented with the silver breastplate.

On April 4, 1792, serjeants of infantry regiments were ordered to be supplied with pikes instead of halberts, which were to be laid aside.

On May 8 officers of grenadier and light infantry companies were ordered to make use of their swords instead of fusils, and the serjeants of grenadiers to carry pikes; but the light infantry serjeants were to keep their fusils. The officers of the flank companies were shortly afterwards ordered to discontinue the use of fusils, together with the cross pouch-belts, the sword to be their only weapon. They still, however, retained the two epaulettes, some few years afterwards to be exchanged for wings.

Towards the end of the century the coats for all ranks were fastened down to the waist, completely hiding the sleeved waistcoat, which afterwards became, for the rank and file, an undress or fatigue garment. The lapels of the officers were to be continued to the waist so as to button over occasionally (making a double-breasted coat in fact), or to fasten close up to the front with hooks and eyes, exposing the blue lapels with their silver buttons and silver-laced buttonholes. It was, however, very customary to wear the coat buttoned across, with the tops of the lapels unbuttoned, and thrown back.

The jacket for the soldiers was single-breasted, having ten buttons and loops of the regimental lace across the chest, arranged by pairs. This lace was slightly altered in pattern, somewhere about this period, from one central blue stripe to two blue stripes, one near each edge.

Officers and men of infantry regiments, excepting those of flank companies, were ordered in 1799 to wear their hair queued, and to be tied a little below the upper part of the collar of the coat, to be ten inches in length, including one inch of hair to appear below the binding.

The cocked hat worn by the men was discontinued in 1800, a cylindrical felt chaco taking its place, ornamented with an oblong brass plate bearing the King's crest, and a red and white tuft fixed in front rising from a black cockade. The officers retained their cocked hats, which they wore sometimes across even with the shoulders, at other times fore and aft.

This head-dress had altered most completely since 1793. It was now a huge affair, of half-moon shape, with the black cockade on one side and a red and white feather issuing from the top; a glance at the illustration (Plate 'H'), representing miniatures of Lieutenant Bent in the uniform of the regiment, will afford the reader a better idea of this remarkable covering than mere words can give. Cocked hats were worn by the officers only, men wearing the chacos, until quite 1811; but before this date the extreme height depicted in the above-mentioned miniatures had been lowered.

Plate 'H' represents two different miniatures of Lieutenant Bent, and gives a good idea of an officer's costume about 1800. The miniature to the right is the full-dress uniform with silver lace loops across the chest in pairs, the huge cocked hat worn straight across. Observe the voluminous black silk neck-cloth showing above the large neck or shirt frill then worn. Lieutenant Bent was an officer of a grenadier company, and wore two epaulettes for this reason. Some years back, when the sword-belt was worn on the right shoulder, the other shoulder supported the pouch-belt used for carrying the fusil ammunition. As a grenadier officer he carried the fusil, hence the extra epaulette, to keep this latter belt in its place. The same arrangement applied also to the light company officers. The epaulettes were of silver, ornamented, for all officers, by an embroidered silver star of regimental pattern upon the strap; above this star an officer of grenadiers wore a gold embroidered grenade.

The oval silver breastplate, same pattern as shown in

Plate 'c,' bore a gilt grenade in the centre, with the number '60,' placed in the middle of the girdle. The costume on the left represents the undress or service coat, having no silver lace, nothing, in fact, beyond the buttons. The cocked hat in this case appears worn fore and aft, the tall feather all white, as was customary for officers of the grenadier companies. The silver buttons were of the pattern shown in Plate 'c,' fig. 3, namely garter and motto encircling the regimental number, with crown over.

In 1796 an order was issued regulating the cocked hats of officers, also ordering officers' swords to have

'a brass guard, pommel and shell, gilt with gold; with the gripe or handle of silver twisted wire. The blade to be straight and made to cut and thrust, conformable to orders given in April 1786. The sword-knot to be crimson and gold in stripes. The gorget to be gilt with gold, with the King's cypher and crown over it, engraved in the middle, and to be worn with a riband and rosette at each end, of the colour of the facing of the regimental clothing.

'Wm. Fawcet,
'Adjt.-General.'

This order affected the regiment, inasmuch as the officers lost their very handsome silver gorgets, so long in use, and had to adopt the new regulation gilt pattern.

Plate 'G' represents a serjeant of the grenadier company, 2nd (now the 1st) Battalion, about the year 1801. His dress embodies the principal changes which had recently taken place; note the new shape, with the large brass plate in front, the jacket also, single-breasted, with plain white tape lace, the only arms being now the sword and the pike, his principal badges of rank consisting of the three chevrons upon his arm, and the crimson and blue worsted sash. Chevrons for non-commissioned officers were authorised in 1802, but these marks of distinction had, it is believed, been in use a year or more already—serjeant-majors to have four, serjeants three, and corporals two on the right arm; the first of silver lace, the second of plain white tape lace, and the third of the regimental lace having two blue stripes.

Plate 'F' shows an officer and a private of a light infantry company of one of the battalions,[1] about the end of the year 1801. By this time wings had taken the place of epaulettes upon the shoulders of officers of the flank companies. The light infantry sabre was hung by slings dependent from a shoulder-belt, ornamented by the oval silver plate, as before described. In the figure of the private note the regimental lace on his jacket, lately changed in pattern from one to two blue stripes.

The troublesome queue was abolished in 1808; the same order directed the hair to be cut short in the neck, and a small sponge added to the soldier's necessaries, for the purpose of frequently washing his head.

The following order was published February 9, 1810: 'Field Officers to wear an epaulette on each shoulder. A Colonel to wear a crown and star (on the strap); Lieutenant-Colonel a crown; Major a star; Captains, Subalterns, and the Quartermaster one epaulette only, on the right shoulder. Officers of the flank companies to wear a wing on each shoulder, with a grenade, or bugle thereon, according to their respective company. The Paymaster and Surgeons to wear a Regimental Coat, but without epaulette or sash.'

These changes were certainly authorised at the above date, but several of them had actually been in use for a

[1] The battalion is uncertain, as some of the light companies had already been equipped as Rifles.

few years previous. The exact pattern of the epaulette, with its strap, varied, together with officers' wings, in almost every regiment. The adjutant wore one upon his right shoulder, and an epaulette strap on his left. The paymaster and surgeons carried their swords from a black leather belt, worn under the coat.

'*General Order. Horse Guards, 24th December*, 1811.

'Officers of infantry are to wear a cap of a pattern similar to that established for the Line; a Regimental Coat similar to that worn by the Private men, but with lapels to button over the breast and body; a grey cloth Great Coat, corresponding in colour with that established for the Line, with a stand up collar, a cape to protect the shoulders, and regimental buttons. In case of regiments employed on foreign service, the officers are to wear grey Pantaloons, or Overalls, with short boots, or with shoes and Gaiters, such as Private men's. When at Court, the Officers of Infantry are to appear in long coats, with cocked hats, as at present.' [With regard to the wearing of overalls, or trousers, garments of this nature had been worn in the West Indies and other foreign stations for very many years; indeed it was generally left to the colonels to decide upon the clothing of the lower limbs, away from home service.]

About the year 1812 a new square silver breastplate for officers was introduced into several battalions (see Plate 'd,' fig. 2), and a few years ago a brass breastplate belonging to a private soldier of the old 1st Battalion was discovered in the bed of the River St. Lawrence near Montreal (fig. 1). Its date may be fixed about 1800 or possibly after, but there is no evidence to show clearly how it got into the position in which it was found.

In 1815 the red coat in the 60th came to an end.

PART II

By MAJOR-GENERAL ASTLEY TERRY

THE GREEN JACKET, DECORATIONS, AND ARMAMENT

1797. THE year 1797 marked the commencement of that great change in the appearance as well as in the title of the regiment which was destined to be carried to completion during the next quarter of a century.

The 5th Battalion, added in that year to the 60th Royal American Regiment, was the first rifle corps in the British Army,[1] and became the model in dress and equipment, as in organisation and discipline, for all subsequent battalions and regiments of riflemen.

No official description of the uniform of the officers at this period is known to exist, and no 'Dress Regulations' were published till twenty-five years later; but sufficient details can be gathered from other sources to show that it was, on the introduction of this arm of the service, as it has been ever since, of a light cavalry pattern.

This dress consisted of a green jacket with scarlet facings, black lace, and three rows of silver buttons in front,[2] silver wings, and a crimson silk sash round the waist;

[1] Two companies of the North York Militia appear to have been dressed in green in 1795, but were neither equipped nor trained as riflemen.

[2] A portrait of Lieut.-Colonel Craufurd (afterwards the commander of the Light Division) shows that officer in a plain jacket. It may be that the lace was not added till 1800, or, what is perhaps more likely, this was a working dress.

green pantaloons with hessian boots worn in 'full dress,' at other times blue-grey trousers without stripes; and a black leather pouch-belt with silver plate, whistle, and chain, carried over the left shoulder. The sword was curved, with black leather scabbard and gilt mountings and fastened by slings to a waist-belt. The head-dress was a helmet of black leather with bearskin crest; a green feather on the left side issuing from a scarlet rosette or cockade,[1] and on the right side a silver bugle fastened to a scarf or turban of green silk. This uniform is shown in the sketch of Lieutenant Wolff in Plate 'H.*'

The dress of the men consisted of a green coat or jacket, sloping away at the back into short tails, a red edging all round, plain red collar and pointed cuffs, red and black wings, and white metal buttons. A white serge waistcoat, blue pantaloons with a narrow red stripe, black gaiters with pointed tops edged with red, black leather cross-belts for bayonet and pouch, the former having an oval brass plate engraved with '60' and a bugle; black chaco, slightly bell-shaped, trimmed with green braid and furnished with a red rosette or cockade, a short green plume and a white metal bugle in front; short rifle with loose black sling.

This dress is described in the warrant of April 9, 1800, and the details are taken from a drawing in the South Kensington Museum Library by Lieut.-Colonel Charles Hamilton Smith, D.A.Q.M.G. at headquarters. It is shown in Plate 'I,' together with that of a rifleman of the 6th Battalion, derived from the same source.

It will be noticed that the uniforms of the rifle companies differed somewhat from that of the 5th Battalion. The coat and wings were similar in shape but entirely

[1] This scarlet cockade, which was worn also on the caps of the men, was probably, like the cross on the pouch-belt, an importation of de Rottenburg's. It was revived, no doubt unconsciously, on the field cap in 1881.

green, with red edging all round, including the collar and top of cuffs, small white frill below collar, and white metal buttons; white pantaloons, short black gaiters with green binding and tassels; straight chaco trimmed with green lace, cord, and tassels, black cockade, and short green plume.

A brown belt was worn over the left shoulder with black cord for powder horn; brown waist-belt with brass snake fastening, brass-mounted sword bayonet, and loose brown sling to rifle.

The 5th Battalion, but not the rifle companies, grew the moustache;[1] both used hair powder; the queues or pigtails were worn straight instead of being clubbed, and, after being reduced in length to seven inches in 1804, were altogether abolished four years later, to the great satisfaction of the whole service.

The serjeants carried rifles instead of pikes, and wore a sash or girdle; and all non-commissioned officers were distinguished by chevrons, which became general in 1802.

1803. In the warrant of April 23, 1803, we find 'In the 5th Battalion of the 60th Regiment and in the 95th Regiment of Foot each serjeant shall have for clothing annually, a jacket the sleeves unlined, a waistcoat with serge sleeves, a pair of pantaloons, and a pair of military shoes; and once in every two years, a cap.

'Each Corporal, Drummer, and Private Man shall have for clothing annually, a jacket lined but not laced, sleeves unlined, a kersey waistcoat with serge sleeves, a pair of blue pantaloons made of cloth the same quality as the

[1] Colonel Holden, in *Colburn's U.S. Magazine* for February 1890, informs us that the Rifle Brigade, when first formed, were also permitted this distinction, and his statement is confirmed by a print (the only contemporary one known) in *The Military Library*, 1800.

jacket, a pair of military shoes; and, once in every two years, a cap as above.

'The men are to be stopped the extraordinary charge of 2*s*. 3*d*. on this clothing in consequence of receiving pantaloons instead of breeches.'

Serjeants had the regimental facings on the great-coat, 1806.
to which in 1806 chevrons were added for all non-commissioned officers. The knapsack, or 'bag' as it was called, was made of brown leather.

A little later the jacket was altered; it now fastened 1808.
to the waist, and had three rows of buttons instead of one, the tails were made wider and looped back, but their former shape was resumed before 1812, when shoulder- 1812.
straps were worn instead of wings, and the chaco, reduced in width at the top, was trimmed as in the rifle companies. Plates 'J' and 'K' show the dress of the battalion early in 1812.

Later in the year a considerable change took place. The light dragoon jacket and helmet worn by the officers were discontinued, and a jacket, with pelisse of hussar pattern heavily laced, and chaco were adopted. This dress, which, without the pelisse, is shown in Plate 'L' with such slight alterations as will be mentioned in their order, was continued till the introduction of the tunic in 1855.

The officers' great-coat or cloak was green, with scarlet collar and cuffs, and had three rows of silver ball buttons down to the waist.

The colour of the officers' trousers[1] and pantaloons of the men of the 5th Battalion was changed to green, but blue continued to be worn by the rifle companies in the West Indies.

The warrant of July 15, 1812, gives the following appointments for rifle corps:—For serjeants: sword and

[1] The officers' *pantaloons* had always been green.

belt, pouch* and belt, sling,* sword-knot, whistle with strap, lock cover,* powder-horn* with lace or strap, copper flask,* and bullet-bag.*

For rank and file, in addition to the articles starred, a waist-belt with sliding carriage; and for buglers, a sword and belt.

1813. In 1813 it occurred to the authorities to dress the red battalions of the regiment in green; and the 7th and 8th Battalions, raised this year, inaugurated the change. Each had two companies equipped as riflemen. It is
1814. true that the annual Army List for 1814 gives both these battalions as dressed in red, but this description is omitted the following year, as if the editor were in doubt on the point. On the other hand, Colonel Hamilton Smith, who was from his official position by far the best authority, states that the 7th Battalion wore 'green with scarlet facings.'[1]

The order was now issued for the change of the red battalions to green, with the title of 'Light Infantry,'
1815. and was carried out in the following year from December 15.

The regiment, therefore, now consisted of a rifle battalion and seven light infantry battalions, all dressed in green.

1816. On April 23, 1816, the following letter was addressed to officers commanding battalions:

'Sir,—I have the honour to acquaint you that a Coat has been lodged at the Clothing Office, Whitehall, belonging to H.R.H. The Duke of York, which has been established by him as the dress uniform for the Officers of the 60th

[1] The number of the *Military Register* for April 13, 1814 (vol. i. p. 88) mentions 'the 7th *or Rifle* Battalion of the 60th Regt.,' and (vol. v. p. 486), 'the 8th Battn. *Rifle*.' Although this nomenclature was incorrect, it clearly indicates a green uniform. Moreover, Colonel Charles Leslie, K.H., in his *Military Journal*, speaking of the 8th Battalion, in which he was then a captain, says: 'On December 12 (1813) we marched through London, to the surprise of the natives seeing so many men in green rifle uniform.'

Regiment. You will therefore be pleased to order that a pattern may be taken wth a view that the Officers belonging to the Battalion under your command may have an opportunity of providing themselves whenever they may require a dress uniform.

'(Sd.) R. A. Darling,

'D.A.G.'

Of course this order did not refer to the Rifle Battalion, as will be seen by the following extract from the *Military Register* for August 28: 'H.R.H. The Commander-in-Chief having at length fully determined upon the dress to be worn in the 60th Regiment (the 5th Battn. excepted), we have been favoured, as all Officers may be, with the sight of a pattern jacket with short skirts, lapels lined with scarlet, scarlet cuffs and collar two buttons on each, two rows of buttons on the fore part, a gold bugle on the skirts; the wings worn are gold chain and bullion; green pantaloons with a figure for dress, and a cap of the regulation pattern conformable to the General Orders (315) June 12.' This chaco, which was 7½ inches deep and had a broad top 11 inches in diameter, was trimmed with black lace; a gilt chain, and a gilt bugle in front with cockade above, a tall green feather and gold cap lines.[1] (See Plate 'M.')

From notes in tailors' books, &c., we learn that 'the wings were of chain gilt, beaded centre plate with silver bugle, mounted on dark green and edged scarlet, bullion fringe 1⅛th inch deep. Epaulettes for Field Officers. Skirt ornaments—gold embroidered bugles on dark green with laurels and "60" between. Except in "Full Dress," blue trousers. A black leather pouch-belt with silver

[1] The chaco of the 7th and 8th Battalions had, like that of the 5th, a green tuft instead of plume, and green cap lines. The 5th Battalion had bronze scales.

ornaments.' An officer of the 7th Battalion is shown in Plate 'N.'

At this time the 1st Battalion was quartered at the Cape of Good Hope, and probably owing to being thus cut off from easy communication with the other battalions, the officers did not adopt the pouch-belt, but wore the silver ornaments on a shoulder-belt to which the sword was attached by slings. No doubt this was more strictly in accordance with the regulations for light infantry.[1]

In full dress the gilt gorget continued to be worn for a time, and a black waist-belt to carry the sword.

The undress of the officers of the light infantry battalions was 'a green double-breasted jacket with skirts, red facings, and gold wings; the rifle companies same as the 5th Battalion and Rifle Brigade.'[2]

1817. An order of May 10, 1817, lays down the uniform of regimental staff officers, which appears to be the same as that given later in the 'Dress Regulations' for 1822, except that the quartermaster was to wear a feather.

1818. The reductions following the end of the war with Napoleon included the 8th, 7th, 6th, and 5th Battalions, which were disbanded in the order given; so that by the middle of 1818 the regiment again consisted of four battalions only, viz. the 2nd, which succeeded the 5th as a rifle battalion, and the 1st, 3rd, and 4th Battalions Light Infantry.

An absurd statement has been put forth that riflemen in the 60th came to an end with the 5th Battalion and commenced anew with the 2nd.[3] It might with equal propriety be urged that the death of the sovereign causes

[1] A drawing by Heath, executed just before the battalion was disbanded, shows this shoulder-belt.

[2] The *Military Register*, vol. x, No. 242.

[3] An ingenious attempt has been made to support this mistake by quoting the latter *only* (dated August 4, 1818) of two documents bearing on the subject, suppressing the one we give. The 5th Battalion was disbanded on July 24. The object of the omission is therefore obvious!

a break in the monarchy! The wording of the following memorandum makes the matter quite clear:

'Most humbly submitted to His Royal Highness The Prince Regent: That in consequence of the 5th or Rifle Battalion of the 60th Regiment *being about to be* disbanded, the 2nd Battalion of that Regiment be clothed, equipped, and trained as a Rifle Corps. In the name and on behalf of His Majesty.

'(Signed) GEORGE P. R.

'July 16, 1818.'

When the 1st and 4th Battalions (Light Infantry) 1819.
were disbanded the 2nd (Rifle) Battalion became the 1st, and the 3rd (Light Infantry) the 2nd Battalion.[1]

In 1820 the officers of the 2nd Battalion were author- 1820.
ised to assume the rifle dress of the 1st Battalion, as is shown by the following extract from General Orders:

'Halifax, May 12, 1820. No. 3.

'Colonel Mackie, Commanding 2nd Battn. 60th Regiment, having shown the Commander of the Forces a letter from the Colonel of that Corps, stating that H.R.H. the Commander-in-Chief had officially approved of the dress of the Officers of both battalions of the 60th being the same, the Commander of the Forces feels himself justified in giving Colonel Mackie authority to act upon that information.'

This order evidently heralded the abolition of the last distinction between the rifle and light infantry battalions.

The first edition of the 'Dress Regulations' was issued in 1822, and gives details of the officers' uniform which 1822.

[1] A drawing by Collins of a private of the 1st Battalion shortly before disbandment shows that the custom of wearing the moustache had extended to the light infantry battalions (Plate 'M*').

had been worn by the Rifle Battalion, with slight alterations (such as the shape of the chaco and change of fur on the pelisse from sable to astracan), since 1812.

We give this description in full:

'*Full Dress and Dress*

'*Jacket*—dark green cloth, hussar style; Prussian collar, full three inches deep, ornamented with black mohair braid; scarlet collar and cuffs, single-breasted, with three rows of silver ball buttons,[1] and Russia braid loops very close all the way down the front; pointed cuff, three inches deep at the point, ornamented with braid; figures on the sleeves, side seams, welts, and hips.

'*Pelisse*—dark green cloth; hussar style; black astracan fur collar, four inches deep, single-breasted, with four rows of black olivets and royal cord loops; black astracan fur cuffs, three inches deep, with an inlet of fur to the sleeve, ornamented with a figure, an edging of two-inch fur entirely round the other parts of the pelisse, figures on the side seams, back, and hips, black neck-lines and tassels.

'*Cap*—black beaver, bell shape; about 7½ inches deep; black sunk glazed top eleven inches in diameter; a black silk 2¾-inch band round the top; a 2-inch ditto round the bottom, and two stripes 1-inch ditto (in an angular direction) on each side; a black lace double centre, communicating by a black bullion loop and button to a bullion rosette at the top; black lines and acorn tassels; bronzed scales and lions' heads; black stamped peak.

'*Tuft*—a round black ball.

'*Pantaloons*—dark green cloth or web.

'*Boots*—hessian.

'*Sabre*—as for Infantry.

[1] From 1824 to 1830 these buttons had 'D.Y.O.R.' on them (Duke of York's Own Rifles).

'*Scabbard*—black leather with gilt mountings.

'*Knot*—plain black leather.

'*Waist-belt*—black patent leather one inch wide, with snake ornament in front, plain rings through which hang two slings of similar width, for rings of scabbard.

'*Pouch*—black patent leather; rounded flap, 4½ inches deep, 5 inches wide at top, 6 at bottom; a bugle in the centre, holes bored.

'*Pouch-belt*—black patent leather, 2½ inches wide, with a plate engraved, and lion's head, whistle and chain.

'*Sash*—crimson silk, patent net with cords and tassels to go twice round and tie.

'*Gloves*—white leather.

'*Cravat*—black silk.

'*Undress*

'*Jacket*—similar to dress, only with a less proportion of trimming, and cuff two inches deep.

'*Pelisse*—not worn.

'*Trousers*—dark green cloth.

'*Boots*—ancle.

'*Cap, Tuft, Sword, Scabbard, Knot, Waist-belt, Pouch and belt, Sash, Gloves, and Cravat*—same as dress and full dress.

'*Forage cap*—plain green cloth, welted black leather peak.

'Field Officers wear in addition,

'*Tache Slings*—three of black patent leather, ½ inch wide, attached to rings of waist-belt, and fastening with loops and buckles to rings of tache.

'*Tache*—plain black patent leather, pocket 9 inches deep, 7½ inches wide at top, 11 at bottom; perfectly plain with three rings at top for tache slings.

'*Spurs*—plated chains.

'*Scabbard*—Steel instead of leather.

'*Great Coat*—Plain blue, single breasted, Prussian collar, and regimental buttons.

'*Cloak*—blue, lined with scarlet shalloon, and of walking length.

'*Regimental Staff Officers*

'*Coat*—single breasted; collars, cuffs, and buttons the same as the uniform of the Regiment, long skirts with white kersimere turnbacks, and bugle skirt ornament; without epaulettes or wings.

'*Cocked Hat*—plain with a black silk loop and button.

'Appointments and other articles of dress same as for other Officers, except the sash, which is not worn. The sword-belt to be worn under the coat.'

1824. On June 25, 1824, the old title of 'The Royal American Regiment' was changed to that of 'The Duke of York's Own Rifle Corps and Light Infantry,' and in the following month to 'The Duke of York's Own Rifle Corps,' the 2nd Battalion being ordered to complete its equipment as a rifle battalion.

At this period the gilt mountings of the sword and scabbard were changed to steel; a dark green drooping cock-tail plume was substituted for the tuft in the chaco[1] (the men being supplied with one of horsehair), trousers with black lace stripes took the place of pantaloons, and the great-coat, which had been changed to blue, now reverted to green with an edging of black braid. A field officer of this date is depicted in Plate 'O.'[2] These

[1] A tuft is given in the Regulations to be worn in undress, but it does not appear to have been used. (See Plate 'P.')

[2] This plate shows the moustache still worn, but it probably ceased this year.

alterations were embodied in the Dress Regulations of 1826, in which also black gloves were ordered. 1826.

Officers of this date are represented in Plate 'P,' and a sergeant of 1829 in Plate 'Q.' The latter shows a jacket 1829. with three rows of white metal buttons, short tails, and a red piping all round; collar and pointed cuffs trimmed with black lace, scale-shaped wings, and black lace chevrons edged with red on both arms. On March 20, 1829, an official memorandum was issued giving the expense of officers' outfits in different regiments, among others:

	£	s.	d.
1st Battalion 60th	73	11	9
2nd Battalion 60th	59	11	0
Rifle Brigade	58	2	6

Now why was the cost in the 1st Battalion 60th so much greater than in the 2nd Battalion and the Rifle Brigade?

The reason was this. When the old 2nd Battalion (afterwards the 1st) succeeded the 5th as 'The Rifle Battalion' in 1818, the officers did not like to discard altogether the gold-laced dress coat given them by the Duke of York, and obtained permission to continue to wear it as a Court dress with the old epaulettes or wings;[1] but the custom gradually died out.

Another circular issued from the Horse Guards on July 7 gives the prices of uniforms charged by the different leading tailors, the following firms being quoted, viz. Messrs. Doland, Jones, Fisher, Anstey, Vernon, Oliphant, Moore, and Tatham.

[1] The Rifle Brigade at one time (*circa* 1820–23) wore at Court a tailed coat with silver epaulettes and gorget. Previously (about 1812–14) they wore a white kersemere waistcoat at mess, with the jacket thrown open.

	£.	s.	d.		£.	s.	d.
The jacket cost from . .	8	0	0	to	8	15	0
Pelisse	11	0	0	,,	13	7	0
Trousers	1	5	0	,,	1	10	0
Chaco, Plume, and Tuft .	5	5	0	,,	6	6	0
Sword, Scabbard, and Knot .	4	0	0	,,	4	10	0
Pouch-belt with bronze[1] ornaments . . .	2	9	6	,,	3	4	0
Black Waist-belt . .		18	0	,,	1	0	0
Sabretasche . . .	1	1	0	,,	1	4	0
Forage Cap and cover .		19	6	,,	1	2	0
Sash	1	15	0	,,	2	5	0
Cloak	3	15	0	,,	8	16	0[!]

1830. In 1830, the plume in the chaco was discontinued and the tuft restored. The chaco was reduced in height to 6 inches, which gave the top a still broader appearance; and a large bronze plate took the place of the lace trimming in front; heavy scale chains were also added.

In December of this year the title of the regiment was changed to 'The King's Royal Rifle Corps,' and the following alterations of dress were sanctioned: black buttons[2] instead of white metal on the jacket, and gambroon trousers without stripes for officers during the summer months. Officers of this period are portrayed in Plates 'R' and 'S.'

During the 'thirties, the undress or shell jacket, which was also worn at mess, had no cord across the front; this

[1] This was probably a suggestion; at any rate none but silver were ever worn in the 60th, though bronze ornaments are shown in a print of an officer R.B., 1830.

[2] Plate 'k'; figs. 1 and 2 give representations of the buttons worn on the jackets of officers and men. The former were of composition, the latter of brass japanned black, a specimen of which was found in 1906 on the site of the old barracks at Nassau in the Bahamas, where a company of the 2nd Battalion was detached from Jamaica in 1841–44.

was introduced by Colonel Trevelyan in 1841, together with a waistcoat of scarlet velvet.

On state occasions pantaloons of black silk web with figured braiding, and hessian boots were still worn—and were not finally discarded till 1856. (Plate 'N.*')

In 1834, when the large chaco plate was discontinued, the cockade was resumed and the cap-lines omitted. 1834.

An officer of this date is shown in Plate 'T.'

A group of the regiment appears in Plate 'U,' and a serjeant and private of 1842 are given in Plate 'V,' which shows the black silk sash worn by the former. 1840.[1]

The broad-topped chaco now disappeared, the issue of 1844 being commonly called the 'Albert Pot.' 1844.

The 1st Battalion always wore cloth facings on the mess jacket, but in 1845 the 2nd Battalion wore velvet. 1845.

The forage-cap was quite plain with rather a wide top, without much stiffness in the sides.

It may here be noted that the distinctive dress of the regimental staff officers laid down in the Regulations of 1822 and 1826 had disappeared by 1831, when 'the uniform of the Regiment without the sash' was ordered. In 1834 the paymaster and quartermaster were directed to wear 'a plain chaco—no tuft,' and in 1846 the medical officers were included in this order.

The staff serjeants wore very handsome jackets with wings, and collar and cuffs heavily laced; the arm badges were elaborate, and, in the case of the quartermaster-serjeant, consisted of a crown, bugle with regimental motto under, and a four-barred chevron, surrounded by a wreath,

[1] Plate 'k,' fig. 8, shows a handsome gilt-mounted button supplied by Messrs. Jennens, who write: 'This was worn about 1840–50. It was the fashion at this period for gentlemen to wear gilt buttons on their coats and vests. Fancy buttons were worn by private gentlemen, but officers of H.M. army wore buttons with the crest or initials of the regiment or corps to which they belonged.'

all embroidered in gold.[1] The badges of a colour-serjeant were also enclosed in a wreath of gold embroidery. Plate
1850. 'X' displays a group of the 2nd Battalion, showing the barrel sash worn by the officers, and also the pelisse exhibited in its various aspects.

1852. Plate 'Y' depicts a corporal and private rifleman
1854. in 1852, and Plate 'Z' a bugler and bandsman in 1854, their jackets being heavily trimmed, the former with black and red, and the latter with black cord.

1855. In 1855 the dress of the whole army was completely changed; coats, jackets, epaulettes, and wings all vanished, and a tunic reigned in their stead. As this dress still exists in the regiment, we will quote the Regulations of 1857:

'*Jacket*—tunic, rifle-green with scarlet collar and cuffs; single-breasted, collar rounded in front. On each side of the breast 5 loops of black square cord, with netted caps and drops, fastening with worked olivets. The top loop eight inches long, the bottom one four inches. A double cord on the shoulders, with small regimental button. The jacket edged all round, except the collar, with black square cord. On the back seams a single cord, forming three eyes at the top, passing under a netted cap at the waist, below which it is doubled, and terminating in a knot at the bottom of the skirt. The skirt nine inches deep for an officer 5ft. 9in. in height, with the variation of $\frac{1}{4}$ inch for every inch of difference in height, and lined with black, and rounded off in front.'[2]

The collar of a field officer was 'laced all round with black lace and figured braiding within the lace.' The sleeve ornaments consisted of 'lace and figured braiding eleven inches deep.'

[1] One of these jackets, which belonged to Quartermaster-Serjeant Dwane, was in possession of the late Major Riley.

[2] Thus, of the new dress as of the old, officers of rifles and hussars wore the same pattern.

The collar of the junior officers was laced round the top only, with figured braiding below for a captain and a plain tracing for a subaltern. The sleeve ornament was an Austrian knot of square cord, a captain having figured and a subaltern plain braiding.

The following badges of rank in black silk were worn on the collar:

Colonels and captains, crown and star; lieut.-cols. and lieuts., crown; majors and ensigns, star.

For undress, a patrol jacket (which still exists) was introduced, corded in front like the tunic, and having scarlet facings, with lace and plain braiding. The 1st Battalion continued to wear the shell jacket before described; the 2nd Battalion had netted caps as on the tunic instead of the side rows of olivets, and the lace on the cuffs differed somewhat. The 3rd and 4th Battalions were again raised about this time, the 3rd following the fashion of the 2nd, and the 4th that of the 1st. The chaco, which was still covered with beaver or silk, now had a straight peak, and the forage cap was trimmed with plain mohair lace, with a figure on the crown. This cap sloped away behind, and the peak was straight. A round cap was worn off parade, and this afterwards superseded the other as regulation.

A grey 'cloak coat' was substituted for the green cloak. A button of this garment is shown in Plate 'k,' fig. 4.

The men's tunic was at first double-breasted, with plain red collar, and cuffs with flaps, the latter also being red. A button is shown in Plate 'k,' fig. 3. After a year or two this was followed by a single-breasted tunic, and shortly afterwards the flap of the cuff became green; and later (1870) a pointed cuff, with scarlet piping only, was substituted. The collar was also laced round the top.

The tunics of the buglers had wings, and were trimmed

with black and red cord. Later on wings were given also to the bandsmen, trimmed with black lace. The buglers carried a short straight sword which, after a time, was supplied also to the band instead of the curved pattern previously worn.

The staff serjeants were given tunics like the officers; their badges, as also those of the colour-serjeants, continued to be of gold, for which black embroidery on red cloth was afterwards substituted, and later still, the wreath was omitted.

1856. Plate 'AA' shows a bandsman in 1856, and Plate 'BB' a private rifleman of two years later.

It was found impossible to obtain a green dye which could be depended upon to keep its colour; the consequence was that the tunics and, more noticeably, the shell jackets worn by the men were of every possible shade from blue to yellow, and presented a most motley appearance on parade. A darker green was therefore introduced which continued to be worn till a permanent dye was obtained about 1897.

By the Regulations for 1857, the medical officers were to wear a shoulder-belt of special pattern instead of the
1861. regimental belt, but in 1861 this order was reversed; they were, however, given a cocked hat with a black plume.[1]

The same year the beaver or silk chaco gave place to one of stitched cloth of smaller pattern, and the badge was reduced in size.

The officers generally wore the sword-belt under the tunic, as it injured the trimming when fastened outside according to regulation. This point was conceded in
1863. 1863, in which year 'scarlet' facings were authorised instead of 'red' for the rank and file.

1870. In 1870 a chaco with black and scarlet braid and a

[1] As a matter of fact they continued generally to wear the chaco.

bronze chain was issued to the men, but was not adopted by the officers. The glengarry cap also came into wear.

In 1873 a busby took the place of the chaco. That 1873.
of the officers was made of black lambskin and the men's of seal. It was trimmed with black cord, and had a cockade or boss above the bronze plate, a black plume with scarlet base and bronze chain. The officers wore black cord cap-lines (Plate 'CC').

The mess-jacket and waistcoat, which had been in use for many years, though mentioned in the Regulations,
was not fully described till December 1876, when the 1876.
following details appeared in General Orders:

'*Shell jacket, 60th Rifles.*—Rifle-green cloth, upright scarlet collar and scarlet cuffs. Black mohair braid, traced, 1¼ inch wide all round the body, forming barrels (or dummies) at bottom of side seams. Side seams trimmed with ¼-inch mohair braid, forming a crowsfoot at each end and in centre. Five waved loops of square cord in front with four rows of knitted olivets.

'Pointed cuffs of 1¼-inch mohair braid with tracing of black Russia, forming a row of small eyes on the outside and inside of the cuffs, and extending 6½ inches from the bottom of each cuff. Collar: mohair braid ⅜ inch wide all round, trimmed through centre with plumes, and row of small eyes along top edge; a loop at bottom of collar to fasten across the neck.'[1]

'*Mess waistcoat.*—Rifle-green cloth, single-breasted, no collar, open half-way down. Hooks and eyes, ½-inch mohair braid on edges, with a ¼-inch braid down the front, one inch from the edge. Scarlet cloth between the two braids, with row of eyes of black Russia down front edge on the scarlet cloth. Pockets trimmed with ¼-inch mohair

[1] On duty the jacket was fastened in front and the pouch-belt worn.

braid, forming a crowsfoot at each end, edged all round with scarlet cloth.'

This dress is shown in Plate 'DD.'

1878. In 1878 the busby was superseded by a helmet made of black felt with bronze binding, spike, chain, and plate.[1]

1881. In 1881 regimental numbers were abolished, and consequently the '60' was eliminated from all clothing, badges, and buttons.[2] Shoulder-straps of black chain gimp were authorised for officers, and the badges of rank, in metal instead of embroidery, were transferred to them from the collar. The shoulder-straps of the men bore the letters 'K.R.R.' instead of the number of the battalion which had for some time superseded the '60.'

A serjeant-major[3] and bugler are shown in Plate 'EE,' and a field officer in Plate 'FF.'

A service forage cap, with flaps to let down and a scarlet boss or cockade with silver bugle, was introduced.

For some years the officers had worn, when on service, a serge patrol jacket very similar to that of the men, and in 1886 this was recognised by the Regulations. It was abolished in 1902.

1890. In 1890 the busby was restored. Though differing in shape from its predecessor, it was made of the same materials, trimmed with black cord and having a black egret plume with a base of scarlet vulture feathers. This has since been changed to a scarlet ostrich plume with a black base. The men, however, still wear the horse-hair plume with the black uppermost.

1902. The round forage cap was discontinued in 1902, and

[1] Cap-lines were sometimes worn by the officers with the helmet.

[2] The men's buttons now had a crown and bugle only, and can hardly be called regimental. (Plate 'k,' fig. 7.)

[3] From 1881 the serjeant-major's badge has consisted of a crown only.

a cap of German pattern was adopted throughout the service with a laced peak for field officers.

A 'service dress' of grey-brown colour was introduced, the different ranks being distinguished by rings of lace and badges on the cuffs. Knickerbockers, putties, and 'Sam Browne' belts were also authorised.

The lace on the collar and cuffs was discontinued on the tunics of field officers and captains, and the handsome old mess dress was discarded for one perfectly plain.

The patrol jacket, which had been adopted by the two rifle regiments in 1856 and afterwards gradually extended to the whole army, was now again confined to its original proprietors. Wonderful to relate, it has undergone no alteration except in colour and the addition of shoulder straps. It may be seen in Plates 'DD' and 'GG.'

Mounted Officers

The sabretasche was worn without ornament from 1797 till 1874,[1] when a silver bugle was authorised to be placed on the flap. It was abolished throughout the service in 1902.

The horse appointments in use in early days are nowhere described, but those of 1824 are shown in the coloured plate of that date (Plate 'O').

They were as follows: Black bridle with silver buckles and bit-bosses, scarlet front and roses, and white collar; black breastplate, scarlet saddle-cloth trimmed with black lace, and bearskin holster covers.

About 1830 the colour of the saddle-cloth was changed to green and the shape to hussar pattern. This also was laced with black, and had an escalloped edging of scarlet cloth; black front and roses on the bridle (Plate 'R').

[1] Although not recognised by the Regulations between 1856 and 1874, it was not discarded.

In 1835 the saddle-cloth resumed its old shape [1] and was trimmed with two rows of black lace showing a light of scarlet cloth between. The white collar or head-stall was retained.

In July 1842 a cross-piece with small rosettes was added to the front of the bridle, and a black horse-hair plume adopted. The head-stall was changed to black leather and lined with scarlet cloth showing an escalloped edging, and a steel chain was carried on the left side. The saddle-cloth was superseded by a shabracque of black lambskin with an escalloped edging of scarlet cloth. This furniture still exists with the following exceptions: The bearskin holster covers were replaced by black patent leather in 1861, and some ten years later steel reins were adopted instead of the single chain, and have in their turn been superseded by a green head-rope. The plume was changed to black and scarlet in 1874.

The silver bit-boss, consisting of the crown and garter with a bugle in the centre, was formerly inscribed 'Sixtieth Rifles,' but in 1881 'The King's Royal Rifle Corps' was substituted.

For many years mounted officers usually had their trousers strapped with leather and wore box spurs, but in 1872 pantaloons, long boots, and hunting spurs were adopted (Plate 'FF').

Straight-necked spurs are worn at mess.

Canada, India, &c.

When the 2nd Battalion was in Canada in 1844–47 the dress of the officers consisted of a frock coat trimmed with

[1] A saddle-cloth which belonged to Colonel Tempest, who left the regiment in 1838, furnished the authority for this description.

black fur, a fur busby with cap-lines, fur gloves, and long boots. (See Plate 'W.')

The men wore a cap with peak and flaps.

The 4th Battalion, on proceeding to that country in 1861, adopted a patrol jacket with black astracan collar, cuffs, and edging, and the usual trimming of cord and lace. Also an astracan busby, somewhat smaller than that formerly worn.

The 1st Battalion in 1867 wore a similar jacket,[1] but a folding astracan cap instead of the busby.

White uniform is worn in India and other hot climates, with a white helmet and pugaree.

A green and scarlet camarband is also authorised.

The Maltese Cross

In 1870 a correspondence appeared in *Notes and Queries* regarding the origin of the Maltese cross as the badge of the regiment.

Several suggestions were made, with more or less approach to probability, but no definite information was obtained.

Captain St. John Mildmay, of the Rifle Brigade, wrote: 'I consider that the present appointments of the 60th are the same as those worn by Hompesch's Regiment. I suggest that Hompesch, who was a Bavarian, adopted the badge of the Malta Cross either from the war medal then given to Bavarian soldiers, and which was of that form, or else because he was a relation of Ferdinand de Hompesch, Knight of Malta, who in 1797 became Grand Master of the Order.'

Owing to the particular form of the first cross adopted

[1] This patrol jacket was frequently worn at home during the winter.

by the regiment, neither of these propositions appears to afford a satisfactory solution.

The Bavarian decoration (1) had the limbs of the cross curved, and, moreover, does not seem to have been conferred till 1814; the badge of the Knights of Malta was

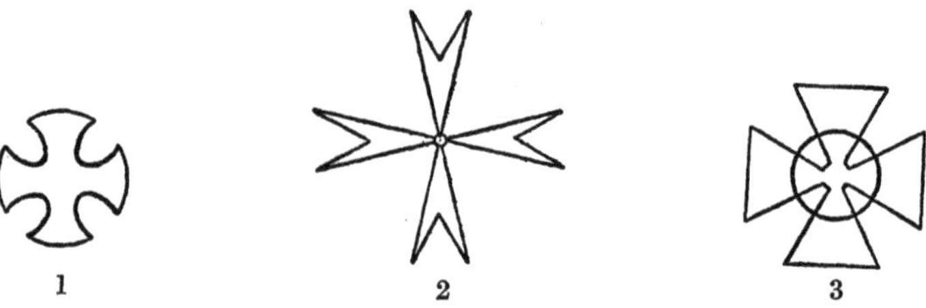

the eight-pointed cross (2); while that of the 60th was a plain cross patée, on the centre of which a circle was placed (3).

As Hompesch was not himself connected with the 60th, it is unnecessary to inquire if this cross was worn by the officers of his light infantry. It is sufficient to know that Lieutenant-Colonel Baron de Rottenburg, who came from that regiment to command the 5th (Rifle) Battalion of the 60th, introduced it as a badge into the British Army. It was afterwards copied,[1] with such differences as necessity or fancy dictated, by light dragoons, the Rifle Brigade, and others.

The Cross on the Pouch-belt

An example of the first cross worn by the regiment is to be seen in the Museum of the Royal United Service Institution on a belt which belonged to Lieutenant
1808. Augustine Evans, who joined the regiment in 1808.

As this belt was worn during the Peninsular war, it is

[1] Captain St. John Mildmay and Colonel Gerald Boyle, old officers of the Rifle Brigade, who have studied the subject, are in agreement with this statement.

quite certain that the cross, in its original condition, must have been perfectly plain, with the exception of the bugle and '60' in the centre. (See Plate 'e,' fig. 1.)

At the end of 1821, when most of the Peninsular honours were granted, Lieutenant Evans found himself serving in the 2nd (Light Infantry) Battalion, and then it was that the rest of the engraving was added. (Fig. 2.) 1821.

This is the cross referred to in the 'Dress Regulations' of 1822 as 'a plate *engraved*.' 1822.

The whistle was carried on chains attached by a ring to a lion's face, and a bugle was fixed on the flap of the pouch. The whole of these ornaments were in silver.

Towards the end of 1822 this plain cross patée was abandoned in favour of the pointed form. Even now, however, it was not the shape used by the Knights of Malta, as, in the latter, the limbs were elongated and the notch or depression at the ends formed a right angle, whereas in this cross the limbs remained equilateral and the depression was comparatively slight.

It was surrounded by a wreath of laurel, and surmounted by a royal crown resting on a bar inscribed 'Peninsula.' The battle honours on the limbs were the same as those on the former cross, but arranged in different order; and the circle, which had a rope edge, contained a bugle only.

This cross is referred to in the 'Dress Regulations' of 1826 as 'a silver plate *embossed*.'

A border or rim was placed round the lion's face, engraved '60th Regiment,' and the form of the bugle on the pouch slightly altered.

It is impossible to give any precise reason for these changes, but it must be remembered that the taste for plainness had died out, and elaborate ornamentation had become fashionable in all details of military dress.

1824. In 1824 a scroll, inscribed in perforation '1st Duke of York's Own Rifle Corps,' was added and placed under the cross.[1]

The belt worn by H.R.H. the Duke of York has these ornaments,[2] which are shown in detail in Plate 'f.' They are also to be seen in Plate 'O.'

1827. In 1827, after the death of the Duke of York, the badge was again altered, and took the form which it still retains. For some forgotten reason the wreath was omitted, the circle was doubled and inscribed with the legend 'The Duke of York's Own First Rifle Corps,' and the '60' was replaced in the centre between the strings of the bugle. Being now no longer required, the scroll was abandoned. (Plate 'g,' fig. 1.)

The border round the lion's face was made more ornamental, and pierced with the motto 'Celer et Audax' which had been restored to the regiment three years previously. (Fig. 2.)

1830. In December 1830, when the title of the regiment was changed to 'The King's Royal Rifle Corps,' the necessary alteration was made on the cross.

1853. By some oversight, when the honours for the Punjab Campaign were granted in 1853, they were allotted to the 1st Battalion only,[3] and a scroll inscribed '1st Battn. Punjaub, Mooltan, Goojerat,' was adopted by that battalion and placed under the cross. The mistake was soon rectified and the names added to the cross, but the scroll remained in use for some years till the old pattern died out.

1884. From time to time honours continued to be added, and

[1] It is not known if the 2nd Battalion had a different scroll or did not wear one at all.

[2] The appointments worn by the Duke of York were presented to the officers of the 1st Battalion by H.M. King George IV.

[3] See *Army List* for May 1853.

in 1884 a slight change was made in their arrangement. The regimental motto, 'Celer et Audax,' was placed on the bar, and 'Peninsula' relegated to its proper position among the other names.[1] (See Plate 'g,' fig. 3.)

On the accession of King Edward VII, the shape of the royal crown was changed to the form generally known as 'heraldic' or 'Gothic.'

In consequence of the accumulation of 'honours' it 1906.
became impossible to find room for them on the cross. His Majesty therefore decided to restore the old 'Peninsula' cross of 1830,[2] with the old crown, and the '60' in the centre, omitting the names of all other campaigns and battles. (See Plate 'h.')

The Cross on the Cap

The original badge worn on the caps of the Rifle Battalion was a bugle—or horn as it was sometimes called—in silver for the officers and white metal for the men.

Between 1815 and 1830 no metal badge was worn by 1815.
the officers, but the men had a bronze bugle in a circle or garter, with a crown above, which they continued to wear till 1844, when the bugle alone was resumed.

The light infantry battalions, 1815–20, wore a bugle in gilt or brass, with the addition, for the men, of the number of the battalion below. (See Plate 'M.*'[3])

[1] When the honours were confined to those granted for the war with France, commonly called 'The Peninsular War,' the old plan was undoubtedly correct, but when other names were added, notably 'Louisbourg' and 'Quebec,' of a prior date, a rearrangement became necessary.

[2] 'Busaco,' granted in 1880, was added.

[3] A duplicate of this plate, with, of course, a difference of 'honours,' was authorised for the Rifle Brigade, the officers of which regiment adopted, at the same time, a cross of similar form, within a wreath, for their pouch-belt. In confirmation of this, Colonel Verner, in a communication to a military journal in 1895,

Note 3 is misplaced, it refers to Plate i on the following page.

1828. In December 1828, when the shape of the chaco was ordered to be changed throughout the service, a plate was designed for the infantry, consisting of a large star surmounted by a crown, and having the badge of each regiment in the centre.

For rifle regiments this plate was of course of bronze and included the Maltese Cross.

This cross differed from that on the pouch-belt in having knobs at the points and lions between the limbs; a scroll at the top was charged 'Peninsula,' and another below 'Celer et Audax.' A wreath was placed on the star round the cross similar to that which had recently been worn on the pouch-belt. (See Plate 'i.')

The cockade or boss was now omitted, as the space it had hitherto occupied was covered by the large crown at the top of the star. This plate was taken into wear
1830. in 1830.

1834. In 1834 the star with crown was discarded, the cockade (with a small crown on it) being resumed, and the cross, within the wreath, placed below it. The wreath was soon afterwards omitted (see Plate 'j,' fig. 1), but this cross, reduced in size in 1861, remained in use till the
1878. introduction of the helmet in 1878.

A large cross was now adopted for all ranks with a crown of proportionate size; the points were without knobs, but the lions between the limbs were retained;

wrote: 'The present cross of R.B., with lions between the limbs and knobs on points, was adopted by R.B. about 1828–29.' Before that date, other patterns had been in use. Sir William Cope, the regimental historian, says: 'The pouch-belt originally had only a whistle and chain affixed to a lion's head. I do not know when the Maltese Cross was first adopted—probably when the names of victories were first granted to the Regiment.' [With the exception of 'Peninsula' and 'Waterloo,' this was in 1821.] 'It was first surmounted by a sitting figure of Fame. An eagle was, I believe, afterwards adopted for a time.' It would be interesting to know if a specimen of either of these curious plates has been preserved.

the scrolls were smaller, and the centre of the circle was perforated, showing the bugle and '60' in relief over scarlet cloth. (Plate 'j,' fig. 2.)

In 1881 the '60' was omitted. 1881.

In 1883 the bronze cap-plate was assimilated in shape 1883.
to the cross on the belt.

On the re-introduction of the busby a small cross again 1890.
became necessary, the crown being, as before, detached and placed on the cockade. This cross is badly proportioned and its outline quite incorrect—a fault due to the ignorance of the War Office draughtsman or some stupid tailor.

Medals for Merit

Three of these medals have come under notice, and are shown in Plate 'l.'

The first is a large silver medal awarded to Edward Thorpe of the 60th Royal American Regiment by Lieut.-Colonel William MacLeod of the 59th Regiment on June 26, 1800.

Thorpe probably belonged to the 4th Battalion, a detachment of which had just been withdrawn from the Leeward Islands, including, no doubt, Antigua, where the 59th Regiment was stationed. The circumstances which led to this presentation are unknown.

The second is a gold medal presented by the Mayor of Cork to Henry E. Chamney, 60th Rifles. This medal, together with the preceding one, was in possession of Mr. Robert Day,[1] of Myrtle Hill House, Cork, who wrote: 'This interesting medal records two events: one, the gallantry of a soldier in a regiment that has ever been

[1] On the death of Mr. Day these medals were acquired by the regiment.

distinguished for individual heroism and bravery; the other, a flood that is still remembered by many citizens of Cork. When it happened two bridges spanned the north channel (now no longer existing); one of these, at the North Gate, largely contributed to the destruction of the other. It was built on three central and two side arches supported on massive buttresses, which in a flood dammed the water up, so that on the memorable second of November it stood as a wall behind it, fretting itself against the piers, and rushing like a waterfall through the narrow arches. Its site is now occupied by an iron bridge of one span. Its companion, St. Patrick's Bridge, at about 11.30 A.M., was destroyed and partly washed away, and some twelve or more persons who had been crossing at the time were drowned. It was in attempting to save one of these, a woman, that Chamney so nobly risked his life by jumping into the roaring current, and for this the Committee, formed for the purpose of aiding and assisting the poor who suffered so much from the effects of the flood, presented him with the medal, which he was permitted to wear upon the right breast, suspended from a dark blue ribbon.'

The 60th were not quartered in Cork at this time, but Chamney may have belonged to the depôt of the 2nd Battalion at Birr, or perhaps was extra-regimentally employed.

The third specimen is in the form of a bronze cross, and was presented as a reward of merit by 'M. Baker, Harrow,' to Harry Hiscock of the 2nd Battalion. It is undated, and there is no further information regarding it. This cross came into the possession of Major-General Astley Terry, who gave it to the regiment with a collection of war medals.

The Armament of the Regiment

When the regiment was raised officers carried swords 1755.
and espontoons,[1] serjeants swords and halberts, and the rank and file muskets and socket bayonets. Officers of grenadiers carried fusils instead of espontoons.

In 1771 officers and serjeants of light companies also 1771.
carried fusils, and in 1784 the issue of this arm was ex- 1784.
tended to serjeants of grenadiers in place of the halbert.

In 1786 the use of the espontoon was altogether 1786.
discontinued.

In 1792 pikes in place of halberts were issued to 1792.
serjeants of battalion companies.

The regiment was furnished at its formation with the 1755.
ordinary flint-lock musket with steel ramrod and socket bayonet, and this arm continued in use until the introduction of the rifle. The weight of the musket in 1800 was 10 lb. 2 oz., and of the bayonet 1 lb. 2 oz.; length of barrel 3 ft. 4. ins; diameter of bore ·753 in.; 14½ bullets to the pound; charge of powder 6 drs. F.G., with three flints to every 60 rounds.

The following exceptional cases are recorded: In 1758.
1758 some rifles were issued to the 1st Battalion under Col. Bouquet,[2] but it is not known how long they remained in use.

Another instance occurred in 1794, when rifles were 1794.
issued to a battalion.

What was the rifle supplied to the 5th Battalion on its 1797.
formation? Rigaud says it was the 'Baker' rifle, which is very probable. Whatever it was, it appears to have

[1] Including field officers, who were not yet mounted.

[2] '*Add. MS.* 21643, "Receipt for rifles, 6th May, Stanwix to Bouquet, 25 May 1758." *Ibid.* 21632, "Bouquet to Forbes, 20 Aug. 1758."'—Fortescue's *History of the British Army*, vol. ii. p. 334.

been superseded by an improved pattern in 1803. This, at any rate, must have been the 'Baker' rifle, as it is certain that that weapon was issued to the battalion before it embarked for the Peninsula.

Marshal Soult, writing to the French Minister for War in 1813, mentions that the 5th Battalion was armed with a short rifle.

Unfortunately, official papers give no clue to a solution, as the following document, which is preserved among the records kept at H.M. Gun Wharf, Portsmouth, will show:

'No. 43. Counterpart.

'Delivered out of His Majesty's Stores at the Tower to Mr. Clarke, Carrier, the undermentioned Arms to be by him conveyed to Portsmouth, and there delivered to William Spencer, Esq., Ordnance Store-keeper, who is to apply to Admiral Gambier to forward them to Nova Scotia, being for service of the 5th Battalion of the 60th Regiment of Foot.

'Nos. 1 to 35	Rifle Musquets with Sword Bayonets, Scabbards, Boxes, Ball-Drawers, Wiping Eyes, and Leavers.	350	In 35 Chests.
	Rifle Musquets without Do.	350	

'(Signed) W. Weaver.

'Office of Ordnance
30th May, 1803.'

'Office of Ordnance, Portsmouth,
'6th June, 1803.

'Delivered into His Majesty's Stores at this Office the Chests above-mentioned.

'(Signed) W. Clark.'

This rifle was very defective, being most difficult to load, and the commanding officer had to apply for mallets to force the bullet into the barrel, and for a better pattern of powder-horn or flask instead of that originally supplied. The rifle was fitted with a long triangular bayonet or sword, which was shortly afterwards superseded by a flat one.

It is probable that the rifle companies were armed with the same weapon, which had, of course, a flint-lock.

The late Captain O'Callaghan, who was an expert in regimental history, and frequently employed as such by the War Office, told the writer that he had proofs of the existence of a rifle company or companies in the regiment prior to the formation of the 5th Battalion. His unfortunate death almost directly afterwards prevented this information being given, but the following document from the records at Portsmouth evidently points to the 4th Battalion having a rifle company, though it is impossible to say if the men were dressed in green till a year or two later:

'Portsmouth.

'This Indenture made the 25th day of January 1798 in the 38th year of the reign of Our Sovereign Lord George the Third, by the Grace of God King of Great Britain, France, and Ireland, Defender of the Faith, &c. Between The Most Noble Marquis Cornwallis, &c., Master General of His Majesty's Ordnance, and the Principal Officers of the same, on the behalf of the King's most Excellent Majesty on the one part; And Mr. Midhurst of the Inspector General's Department on the other part; Witnesseth, That the said Mr. Midhurst hath received out of His Majesty's Stores, within the Office of Ordnance at Portsmouth, the particulars undermentioned for use of the

Riflemen ordered to St. Domingo on their way to Jamaica. In pursuance of the Board's Order dated 16th instant; viz.:—

Pigs of lead—9—weight—lbs. 1538
Carbine Flints . . . 2000.

'(Signed) W. Midhurst,
'*Asst. Inspector General.*'

Improvements to this rifle[1] were made from time to
1841. time, and it continued in use in the regiment till 1841, when the 'Brunswick' percussion rifle was issued.

This rifle had two deep grooves, making a full turn, and a belted ball, and was fitted with a sword bayonet. It was about the same length as its predecessor, but the sword was longer—22 inches. It had the same great defect, namely difficulty in loading, although two notches were cut in the muzzle to ensure the belt of the bullet dropping into the grooves.

During the Kafir war it was found necessary to furnish each rifleman with a packet of plain musket ammunition, to be used when easy and rapid loading was essential.

1855. When the 3rd Battalion was raised in 1855, the men were armed with the long 'Enfield' rifle and bayonet: three grooves, one turn in 78 inches, diameter of bore ·577 inches.

1856. In 1856 the short 'Enfield Pritchett' with sword was
1857. issued to the 2nd Battalion, and in 1857 the long one to the 1st and 4th Battalions.

1858. In 1858 the long rifles in possession of the 1st and 3rd Battalions were exchanged for short ones.

The 'navy rifle' was issued to the 4th Battalion in

[1] A General Order dated Quebec, September 1818, directs the smooth-bore arms in possession of the 2nd Battalion to be handed over to the storekeeper, rifles having been received by that battalion.

1859 and to the 1st Battalion in 1861. This rifle had five grooves with one turn in 48 inches. 1859–61.

The 'Whitworth' rifle—one turn in 20 inches—was issued to the 1st and 2nd Battalions in 1864. 1864.

The 'Snider' breech-loader[1] was supplied to the 2nd and 4th Battalions in 1866, to the 1st Battalion in 1867, 1866–67.
and to the 3rd Battalion in 1871. 1871.

The 'Martini-Henry' rifle was issued to the 1st Battalion in 1873, to the 3rd and 4th Battalions in 1874, 1873–4.
and to the 2nd Battalion in 1877. The barrel of this 1877.
rifle had seven grooves with one turn in 22 inches.

In 1889 magazine rifles were introduced, the first 1889.
being the 'Lee-Metford' with seven grooves, one turn in 10 inches. This, with slight alterations, remained in use till the 'Lee-Enfield' took its place, and in 1906 the barrel—with five grooves—was reduced in length from 2 ft. 6 in. to 2 ft. 1 in. This rifle has an adjustable barley-corn foresight, radial backsight, with vertical adjustment and wind-gauge.

A short flat bayonet has been in use since the introduction of the magazine rifle.

[1] This was a converted 'Enfield.'

PART III

By S. M. MILNE

THE COLOURS

The use of colours had been established for quite two centuries before the regiment was raised, and, in the British service, had dropped from one for every company to two for each regiment or battalion. The earliest regulation regarding colours was issued September 14, 1743, and ordered that 'the first Colour of every marching regiment of foot is to be the great Union, the second Colour to be of the hue of the facing of the regiment, with the Union in the upper canton, except those regiments faced with white or red, when the second Colour is to be the red cross of St. George on a white field, and a small Union in the upper canton. In the centre of each Colour is to be painted in gold Roman figures the number of the rank of the regiment within a wreath of roses and thistles on one stalk, excepting those regiments which are authorised to wear royal devices, or ancient badges, when the number of the regiment is to be painted towards the upper corner.'

The great Union above-mentioned consisted of the two crosses of St. George and St. Andrew combined upon a blue ground. Not for some sixty years after did the St. Patrick's cross appear upon the Union. A further royal warrant was issued in 1743, followed by another in 1751, the latter differed little from its first predecessor of 1743, except that it officially named the first as the King's Colour, and the second the Regimental Colour. The dimensions now authorised for the length of the pike (spear and ferule included) were 9 ft. 10 in., width of flag 6 ft. 6 in., depth on the pole 6 ft. 2 in.

In December 1768 a royal warrant was issued detailing the badges of two royal regiments, the 42nd and the 60th, and referring to the 60th as follows: 'The sixtieth or Royal Americans, in the centre of their Colours the King's cipher within the garter with crown over, in the three corners of the second Colour, the King's cipher surmounted with a crown.'

The regiment having been raised in 1755, there can be no doubt that these Royal badges had been granted, and were fully displayed upon the earliest stand of colours. (Plate 'a.') The first number was '62,' but very soon altered to '60.'

A considerable amount of information with regard to colours may be gleaned from a careful search in the old 'Inspection Returns' preserved in the Public Record Office. The earliest to be found about the 60th is dated December 19, 1783, and refers to the 1st Battalion, stationed at Spanish Town, Jamaica. The following is noted by the inspecting officer: 'The battalion has one Colour good, and one in bad condition; having been with the regiment ever since it was raised, they are worn'; adding 'The officers are armed with swords and fusils. The battalion, excepting thirty-eight Germans, who are valuable soldiers and have served His Majesty long and faithfully, may now be considered as much British as any corps in the service.'

Of the same battalion, the 1st, on the occasion of its inspection at Montreal, September 19, 1787, the colours were reported as 'in bad condition.'

Again at Montreal, October 10, 1790, inspected by Colonel Buckeridge, 65th Regiment, the colours were reported upon as 'good.' Therefore, at some date between 1787 and 1790, the 1st Battalion must have had new colours, being the second stand it received, and of precisely the same design as the first, except as regards the number.

On October 25 the same year, the 2nd Battalion was inspected at Sauteau Recollett by the same officer, who stated that it 'had new Colours in 1787'—doubtless the second stand this battalion had possessed. The 3rd Battalion, inspected at Dominica, May 31, 1790, had received new colours in 1788.

The 4th Battalion, reviewed by General Matthews at Barbadoes, May 19, 1788, had 'Colours in good condition'; the 3rd and 4th Battalions had not long been raised (for the third time). On September 2, 1792, the 1st Battalion was inspected at Longueuil by His Royal Highness Prince Edward (afterwards Duke of Kent).

At Halifax in 1804, General Bowyer, on inspecting the 5th Rifle Battalion, gives the number of rifles carried by the men, and reports upon the high character and condition of the battalion. Of course no mention is made of colours: as a rifle corps it never had any.

One cannot omit mentioning the few remarks made by General F. Maitland upon the occasion of his inspection, July 5, 1805, if only for the pathetic conclusion. After remarking of the 2nd Battalion that the colours were in good order, General Maitland proceeded to say: 'Nearly all the captains and subaltern officers at present serving with the battalion are foreigners who came from Hompeschs' or Lowenstein's corps; these have been effective for eight or nine years, and have been the stand-by of the battalion; they are industrious and good officers, and deserve notice for their good conduct, and want of friends.' Promotion evidently must have been very slow at that time.

The colours of the 1st Battalion were reported upon as being in bad order, April 18, 1807, at Maroon Town, whilst those of the 6th Battalion, April 1, 1807, at Port Antonio, seemed to be in good order. Of the colours of the senior four battalions inspected during the year 1808, those of the 1st Battalion were bad, as last described; the others in good order.

In the numerous changes of various kinds which took place from 1815 to 1819, it is quite impossible to follow the colours, either to make out their history, or their fate.

The very last notice may be found in the first half-yearly inspection of 1820 of the 2nd Light Infantry Battalion, clearly showing that this battalion still possessed colours, for the Inspecting Officer reports them in bad condition. This 2nd Battalion became in 1824 a rifle corps like the 1st Battalion, but possibly before this occurred the colours may have been retired or gone out of use, for they are never mentioned again.

The only specimen of a colour which the author of these notes has met with, is one which was given to the late Captain Holbech about 1844, whilst serving with the regiment in Canada, by a descendant of Colonel Andrews, who commanded one of the battalions (latterly the 1st), 1815–23.[1]

The coloured illustration represents these colours as they would appear when new; notice the shamrock in the Union wreath, also St. Patrick's cross added. There is nothing upon them to distinguish the battalion after the Union in 1801, simply the number of the regiment on

[1] A colour of the 4th Battalion, given to the Colonel-Commandant, Augustine Prevost, on the disbandment in 1783, is now in the possession of his great-grandson, Sir C. Prevost.

the dexter canton, or upper corner. As may be noticed by comparison with the old 1755 colours they are in strict accordance with the Regulations. (Plate 'a*.') It will be at once noticeable that the regimental colour has the Union wreath embroidered round the centre, whereas the King's colour is quite plain. One can only remark that such was not infrequently the case with royal regiments; at all events a little latitude in this respect was tolerated.

NOTE.—When the red battalions were put into green, in 1815, their colours may have been appropriated in the usual way by the colonels, or they may have remained on charge, as Mr. Milne has shown was the case with the 2nd Battalion in 1820. It is most improbable, however, that they were carried on parade. A battalion dressed in green with blue colours would have appeared grotesque indeed. Moreover, the 7th and 8th Battalions, raised in 1813, do not seem to have received any. The Rifle Battalion of course had none.

A. T.

THE END

www.ingramcontent.com/pod-product-compliance
Ingram Content Group UK Ltd.
Pitfield, Milton Keynes, MK11 3LW, UK
UKHW021840270726
14058UKWH00002B/263

9 781843 424574